The Stock Market

Understanding and Applying Ratios, Decimals, Fractions, and Percentages

Orli Zuravicky

PowerMath™

The Rosen Publishing Group's
PowerKids Press™
New York

Published in 2005 by The Rosen Publishing Group, Inc.
29 East 21st Street, New York, NY 10010

Book Design: Michael J. Flynn

Photo Credits: Cover © Doug Mazell/Index Stock Imagery; pp. 5, 27 © Bettmann/Corbis; p. 7 ©
Ed Eckstein/Corbis; p. 8 by Maura B. McConnell; pp. 10–11 © Peter Gregoire/Index Stock Imagery;
pp. 12, 22 by Michael J. Flynn; p. 14 © William Taufic/Corbis; p. 18 © Powerstock/Index Stock Imagery;
pp. 20–21 © RO-MA Stock/Index Stock Imagery; p. 24 © Ben Mangor/SuperStock; p. 29 ©
Vittoriano Rastelli/Corbis; p. 30 © Strauss/Curtis/Corbis.

Zuravicky, Orli.
  The stock market : understanding and applying ratios, decimals, fractions, and percentages / Orli Zuravicky.
      p. cm. — (PowerMath)
  Includes index.
  ISBN 1-4042-2929-9 (library binding)
  ISBN 1-4042-5121-9 (pbk.)
  6-pack ISBN: 1-4042-5122-7
  1. Fractions—Juvenile literature. 2. Stock exchanges—Juvenile literature. I. Title. II. Series.
  QA117.Z87 2005
  513.2'4—dc22
                                    2004003250

Manufactured in the United States of America

Contents

What Is the Stock Market?

Have you ever heard a TV news reporter mention the stock market? The stock market might seem like a difficult topic, but it's an important one to understand. This book will teach you about the stock market with the help of a few simple math skills.

First, what is a stock? A stock is a piece of a business that people can buy. A person who buys stock in a company is called a stockholder. Owning stock in a company means that you own a part of that company. When you and other stockholders buy stock, you pay for everything in the company, from pencils and rugs to salaries and buildings. Companies sell stock because they know they can grow and succeed with the help of stockholders' money. People **invest** in stocks because they hope to make even more money than they invest. Billions of dollars of stocks are bought and sold every day.

How do people make money with stocks? The simplest explanation is that when a company grows, its success puts it in a higher **demand** with the public. A stock's worth rises when more people want to buy the company's products and stock. Stockholders purchase stock for a certain price. If they still own the stock when the price of the stock rises, they have made money.

> This certificate from 1903 shows Henry Ford's stock in his own automobile manufacturing company. With many companies now relying on computers, paper certificates are often not distributed anymore.

INCORPORATED UNDER THE LAWS OF THE STATE OF MICHIGAN.

NUMBER

SHARES 255

FORD MOTOR COMPANY.

CAPITAL $150,000.

FULLY PAID. NON-ASSESSABLE.

THIS CERTIFIES THAT _Henry Ford_ is the owner of _Two Hundred and Fifty Five_ Shares of the Capital Stock of

FORD MOTOR COMPANY,

transferable only on the books of the Corporation by the holder hereof in person or by Attorney upon surrender of this Certificate properly endorsed.

In Witness Whereof, the said Corporation has caused this Certificate to be signed by its duly authorized officers and to be sealed with the Seal of the Corporation this _26th_ day of _June_ A.D. 190 3

GREGORY, MAYER & THOM CO., DETROIT.

Shares $100 Each.

When you think of a market, you probably picture a place where certain services or products are sold. The stock market is a market where stocks are bought and sold in a process called trading. However, the stock market isn't a single place. It is actually a **network** of different markets that are all connected.

The first official place to trade stocks in America began in 1792 under a tree on Wall Street in New York City. This market is known today as the New York Stock Exchange, or NYSE. As years passed, the country grew and stock exchanges opened in other cities. Soon, advancements in **technology** made it possible to trade stocks over the phone and on the computer.

Today, 3 major exchanges exist in the United States: the NYSE, the American Stock Exchange (AMEX), and the National Association of Securities Dealers (NASDAQ). The NYSE and AMEX are called traditional markets because they are physical places where trading occurs. The NASDAQ has no physical market location—stocks are traded through computers and telephones. Companies' stocks are listed in these different markets based on certain requirements. All of these markets, smaller local exchanges, and international stock exchanges make up what we call the stock market.

The American Stock Exchange, located in New York City, began in the 1800s with some men selling stocks on the streets. In 1921, the stock exchange moved indoors and became a major trading market.

Stock Math

To understand stocks and the stock market, you need to understand fractions, **ratios**, **decimals**, and percentages. These types of numbers are used every day to explain information about the stock market.

A fraction is a piece of a whole. If you divide a whole pizza into 10 slices and eat 1 slice, you have eaten a fraction of the whole pizza. In fraction form, the amount you have eaten is written $\frac{1}{10}$. The top and bottom numbers of a fraction have names. In the fraction $\frac{1}{10}$, the **numerator**, 1, is the number of equally sized pieces you have eaten. The **denominator**, 10, is the total number of equally sized pieces that makes up the whole.

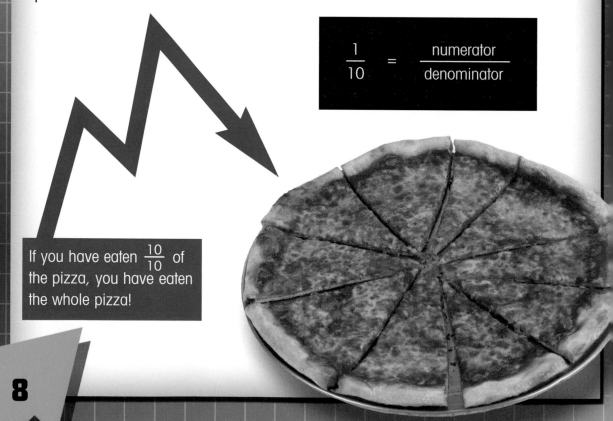

$$\frac{1}{10} = \frac{\text{numerator}}{\text{denominator}}$$

If you have eaten $\frac{10}{10}$ of the pizza, you have eaten the whole pizza!

You could think of 1 stock as 1 piece of pizza. Since a stock is 1 part of a company that has many parts for sale, a person who owns stock owns a fraction of a whole company. Let's say that you own stock in—or a share of—Maggie's Pizzeria. You can use fractions to find out how much of Maggie's Pizzeria you own with your shares of stock. If you own 100 shares out of a possible 1,000 shares, what fraction of Maggie's Pizzeria stock do you own? The numerator is the number that you own. The denominator is the total number of shares available.

$$\frac{100}{1,000} = \frac{\text{your shares}}{\text{total shares}}$$

You own $\frac{100}{1,000}$ of Maggie's Pizzeria.

Fractions can be reduced, or simplified, if the same number can divide both the numerator and the denominator without leaving a remainder. The highest number that can evenly divide both the numerator and the denominator of the fraction $\frac{100}{1,000}$ is 100.

$$\frac{100}{1,000}$$

$$\begin{array}{r} 1 \\ 100\overline{)\,100} \\ -100 \\ \hline 0 \end{array}$$

$$\begin{array}{r} 10 \\ 100\overline{)\,1,000} \\ -1\,00 \\ \hline 00 \end{array}$$

$$\frac{100}{1,000} = \frac{1}{10}$$

The reduced fraction is $\frac{1}{10}$. If you own 100 shares of stock in Maggie's Pizzeria, you own $\frac{1}{10}$ of the entire company.

Another type of number used in stock information is a ratio. A ratio compares amounts. If you put mushrooms on 5 slices of an 8-slice pizza, the ratio of mushroom slices to the number of plain slices of pizza is $5:3$. When we say this ratio aloud, we say "5 to 3." The ratio of plain slices to the mushroom slices is $3:5$. A ratio can be written in fraction form, $\frac{3}{5}$, or with a colon, $3:5$.

Let's see how this idea applies to owning stock. You own 100 shares of stock in Maggie's Pizzeria and 3,000 shares of stock in Universal Pizza Company. What is the ratio of stock you own in Maggie's Pizzeria to stock you own in Universal Pizza Company?

In some big cities and on television stations, you may see information on stocks; including fractions, scroll across a screen like this one.

The ratio of stock in Maggie's Pizzeria to
stock in Universal Pizza Company

"100 to 3,000"
or
100 : 3,000
or
$$\frac{100}{3,000}$$

We can reduce ratios the same way we reduce fractions. The highest number that can evenly divide both the numerator and the denominator of the ratio $\frac{100}{3,000}$ is 100.

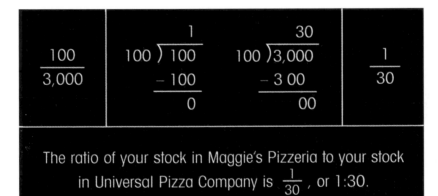

The ratio of your stock in Maggie's Pizzeria to your stock in Universal Pizza Company is $\frac{1}{30}$, or 1:30.

The ratio 1:30 means that for every share of stock in Maggie's Pizzeria that you own, you own 30 shares of stock in Universal Pizza Company.

Reading the Stock Tables

Information on the stock market is **computed**, organized, and printed in stock tables. A combination of numbers and letters is listed to tell people how different stocks are doing in the market. One number listed in the stock tables is the P/E ratio (or the price per earnings ratio). This is the ratio of a stock's price per share to its company's earnings per share over a period of time. This allows people to see how their stock price compares to the company's earnings and how their stock compares to other stocks.

For example, if a share of stock in The Candy Company is priced at $4 and the company earns $1 per share, the P/E ratio is 4:1, or $\frac{4}{1}$. This P/E ratio would be listed as the whole number 4 because the numerator can be divided evenly by the denominator. In other words, every $4 invested produces $1 in earnings.

$$\frac{\$4}{\$1} = \frac{\text{price per share}}{\text{earnings per share}}$$

The P/E ratio is $\frac{4}{1}$, or 4.

The stock tables may look complicated, but each number tells investors information that can persuade them to invest more money or sell their stock.

The P/E ratio is often used to compare companies that have similar products. Companies that are expected to grow in the future—and thus have higher earnings—usually have a higher P/E ratio than companies that are losing value. Therefore, the higher the P/E, the more likely investors are to buy that stock and the more likely it is the stock will become more valuable. For example, The Candy Company P/E ratio of 4 will help you see how your stock is doing in relation to the stock in other candy companies.

If a share of stock in The Sweets Company costs $56 per share and the company makes $10 per share, what is the P/E ratio? Which company has a higher P/E ratio, The Candy Company or The Sweets Company?

$$\frac{\$56}{\$10} = \frac{\text{cost per share}}{\text{company's earnings per share}}$$

$$\begin{array}{r} 5.6 \\ 10\overline{)56.0} \\ -\ 50 \\ \hline 60 \\ -\ 60 \\ \hline 0 \end{array}$$

The Sweets Company has a P/E ratio of 5.6, which is higher than The Candy Company's P/E ratio of 4.

The numbers in stock tables help people compare the stocks of companies.

14

Decimals are also used to relay stock information. Decimals, like fractions, can show a number that is not whole. The decimal point is a small dot that separates whole numbers from fractions of whole numbers. The number to the left of the decimal point is a whole number. The number to the right of the decimal point is a fraction. Similar to the tens, hundreds, and thousands places of whole numbers, the places to the right of the decimal are named the tenths, hundredths, and thousandths places. For example, look at the number 3.5. The number 3 is a whole number. The number 5 is a fraction of a whole number. We say this number as "three point five" or "three and five-tenths."

Decimals can represent cents, which are fractions of dollars. If you bought stock in The Telephone Company for $\frac{17}{4}$ per share, how much money did you spend on each share? To get the decimal form, divide the numerator, 17, by the denominator, 4.

$$
\frac{17}{4} = 4 \overline{)\begin{array}{r} 4.25 \\ 17.00 \end{array}}
$$

$$
\begin{array}{r}
- 16 \\
\hline
1\,0 \\
- 8 \\
\hline
20 \\
- 20 \\
\hline
0
\end{array}
$$

You bought the stock
for $4.25 a share.

Let's say you bought stock for $10 a share, and its value went down $\frac{7}{8}$ per share. How much money did you lose per share? To find out, divide 7 by 8. This is the value each share lost in cents. To find out how much each share is now worth, subtract that decimal from $10.

$$\frac{7}{8} = 8\,\overline{)7.000} \quad .875$$

$$\begin{array}{r} .875 \\ 8\,\overline{)7.000} \\ -6\,4 \\ \hline 60 \\ -56 \\ \hline 40 \\ -40 \\ \hline 0 \end{array}$$

Each share is worth $87\frac{1}{2}$ cents less.

$$\begin{array}{r} \$10.000 \\ -\ .875 \\ \hline \$9.125 \end{array}$$

Rounding to the nearest hundredth, the $10 stock is now worth about $9.13.

Information about changes in stock values was once given only in the form of fractions. The U.S. Securities and Exchange Commission, the federal agency that oversees the nation's stock exchanges, ordered all U.S. stock exchanges to convert to decimals in 2001. Let's compare using fractions and decimals with a share of stock worth $5.50. You read in the newspaper that the stock's value went up $\frac{19}{4}$ or 4.75. How much money is the stock worth now? Which method do you prefer, using fractions or decimals?

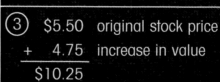

Decimal Method

$5.50   original stock price

$+\ \underline{\ 4.75}$ increase in value

$10.25

The stock is worth $10.25.

Fraction Method

① $5.50   original stock price

$+\ \ \frac{19}{4}$ increase in value

②
$$\begin{array}{r} 4.75 \\ 4\,\overline{)19.00} \\ -16 \\ \hline 30 \\ -28 \\ \hline 20 \\ -20 \\ \hline 0 \end{array}$$

③ $5.50   original stock price

$+\ \underline{\ 4.75}$ increase in value

$10.25

The stock is worth $10.25.

$$\frac{3}{10} = 10\overline{)3.0}\quad\begin{array}{r}.3\\\hline\end{array}$$

$$10\overline{)\,3.0}$$
$$\underline{-\,30}$$
$$0$$

$$\begin{array}{r}100\\ \times\quad.3\\ \hline 30.0\end{array}$$

You have eaten 30% of the whole pizza.

Another way to write a decimal or a fraction is as a percent. Percentages show how certain numbers compare to 100. The highest possible amount you can have without exceeding your original amount is 100%. Referring back to our original example, a whole pizza, or $\frac{10}{10}$, equals 100% of the pizza. If you eat 3 pieces of pizza, you have eaten $\frac{3}{10}$, or .3, of the whole pizza. What percentage of the pizza have you eaten? To find the answer, multiply the decimal number by 100, which shifts the decimal point 2 places to the right.

You can also change percents into fractions. If you eat 30% of a 10-slice pizza, what fraction of the pizza have you eaten? First, we change the percent into a decimal using division. To create a fraction, figure out what place the decimal number is in (tenths, hundredths, and so on).

$$\frac{30\%}{100\%} = \frac{\text{pizza eaten}}{\text{total pizza}} \qquad \begin{array}{r}.3\\ 100\overline{)30.0}\\ \underline{-\,30\;0}\\ 0\end{array}$$

$$.3 = 3 \text{ tenths} = \frac{3}{10}$$

You have eaten $\frac{3}{10}$ of the pizza.

4.94	1855.89	I+	14.31I	95
6.50	297.00	I-	0.40I	9
1.00	301.00	I+	0.50I	21
3.00	493.50	I+	1.50I	8
3.20	513.50	I+	0.30I	4
3.00	636.00	I+	5.00I	7
9.00	289.00	I	0.00I	9
3.00	815.00	I-	2.00I	11
6.00	338.00	I+	5.00I	12
1.00	753.00	I+	8.00I	25
5.00	416.50	I+	2.50I	9
4.50	476.00	I+	3.50I	17
5.00	427.00	I+	4.50I	18
2.00	587.00	I+	14.00I	20
8.00	458.00	I+	2.00I	2
1.00	832.50	I+	6.50I	16
6.00	177.00	I+	0.70I	15
6.00	596.50	I+	10.00I	24

Deutscher Aktienindex   18.05.90   Letzter Wert   1855.89
(1841.58)   +17.36   +14.54

DAX Frankfurter Wertpapierbörse

1870
1861
1854
1847
1840
(VT)   11:20   12:10   13:00   13:45
Höchstwert 10:53 1861.09   Tiefstwert 10:35 1854.94

This stock exchange in Frankfurt, Germany, has a large line graph which shows changes in the amount of stock trading.

People often use percentages to explain changes in a stock's worth. Let's say you originally bought 10 shares of stock for $10 each, for a total of $100. Then your stock goes

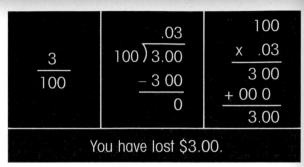

$$\frac{3}{100}$$

$$\begin{array}{r} .03 \\ 100\overline{)3.00} \\ -3\ 00 \\ \hline 0 \end{array}$$

$$\begin{array}{r} 100 \\ \times\ .03 \\ \hline 3\ 00 \\ +\ 00\ 0 \\ \hline 3.00 \end{array}$$

You have lost $3.00.

down 3% in value. How much money have you lost? To find out, first divide 3 by 100 to get the decimal form. Now, multiply the total price you paid by the decimal to find out how much you have lost.

Another number listed in the stock tables is the dividend yield. A dividend is an amount of a company's profits for that year offered to stockholders as a reward for investing their money. The dividend yield is the percentage of the original price paid for the stock that is returned to the stockholder through the dividend.

$$\frac{\$2.25}{\$75} = \frac{\text{dividend per share}}{\text{price per share}}$$

$$\begin{array}{r} .03 \\ 75\overline{)2.25} \\ -\ 2\ 25 \\ \hline 0 \end{array}$$

$$\begin{array}{r} 100 \\ \times\ .03 \\ \hline 3\ 00 \\ +\ 00\ 0 \\ \hline 3.00 \end{array}$$

The dividend yield is 3%. The stockholder receives 3% of the original price per share paid.

For example, let's say the Everything Pets Corporation offers their stockholders a $2.25 dividend on every share of stock. We would calculate the dividend yield for a stockholder who paid $75 per share for stock in the Everything Pets Corporation by dividing the dividend per share by the original price paid per share. Then, convert the decimal into a percent.

Stocks are described in different ways. Two terms often mentioned are "growth stocks" and "blue-chip stocks." Growth stocks are stocks in companies that are trying to grow and are expected to do well. Because they are trying to grow, most of the money they earn is put back into the company instead of providing stockholders with a high dividend yield. Blue-chip stocks, on the other hand, are offered by very large, stable companies that consistently make a profit. These stocks generally offer dividend yields. Why would people choose one type of stock over the other? Some stocks—like blue-chip stocks—are considered safer ways to make money slowly. Others—like growth stocks—are riskier, but could earn you a lot of money quickly.

If you buy shares in a blue-chip company for $20 a share and the company offers you a dividend of $4 a share at the end of the year, what is the dividend yield? Look at the box below to help you find the answer.

$$\frac{\$4}{\$20} \qquad 20\overline{)4.0} \qquad \begin{array}{r} .2 \\ \hline \end{array}$$

$$20\overline{\smash{)}4.0} \\ \underline{-\ 4\ 0} \\ 0$$

$$\begin{array}{r} 100 \\ \times\quad .2 \\ \hline 20.0 \end{array}$$

Your dividend yield is 20%.

The stock market is sometimes referred to as a bear or bull market. A bear market is a market in which prices of stocks are expected to fall. A bull market is a market in which prices are expected to rise. This statue of a bull is located on Wall Street in New York City.

21

Rising and Falling Stock Prices

Market prices are based on public demand. Let's say you want to buy your favorite movie on DVD. If you are the only person who wants to buy that DVD and the seller has 100 copies, they will probably sell it to you for a low price. However, if you are one of 100 people who want that DVD and the seller has only 1 copy, he will sell it to the highest bidder. The same is true for stock market prices. If public demand is rising for a company's products, more people will invest in that stock, and the price of that stock will also rise. If public demand for a company's product is dropping, the price of the stock will also drop. If you see your stock in a company has dropped 10 points, that means the price of your stock has dropped $10 per share.

Public demand for stock is based on many things, including how stable a company is and whether the public thinks that company's product will be successful. Stock prices change from moment to moment because so many stocks all over the world are being bought and sold at the same time.

Plus and minus signs appear before some numbers on this stock table. A plus sign means the stock has gained value. A minus sign means the stock has lost value.

GAINERS ($2 OR MORE)

Name	Last	Chg	%Chg
VasoAcPh n	33.15	+11.04	+49.9
UtdAHlth	5.32	+1.02	+23.7
BluCoat	38.27	+6.75	+21.4
Immersn	8.68	+1.43	+19.7
IndSrvAm	14.67	+2.40	+19.6

LOSERS ($2 OR MORE)

Name	Last	Chg	%Chg
CmpDyn	9.30	-2.40	-20.5
GranCF un	5.61	-1.39	-19.9
Z-TelTch	3.80	-.80	-17.4
WrlssFc	13.50	-2.53	-15.8
Consulier	3.54	-.56	-13.7

Let's say that you bought 100 shares of stock in Planet Music at $10 a share. You have paid a total of $1,000. A month later, the price for a share of stock in Planet Music has gone up to $16. If you sell your shares for the market price of $16, you will get $1,600. If you subtract the $1,000 that you paid for them originally, your profit is $600.

100	shares		100	shares		$1,600	new stock worth
x	$10	price per share	x	$16	price per share	−$1,000	original stock worth
$1,000	original stock worth		$1,600	new stock worth		$600	your profit

You have earned $600 from your original investment of $1,000.

If the share price for Planet Music drops to $6, you will lose money by selling while the price is low. How much money will you lose? If you sell 100 shares at $6 each, you will receive only $600, which is $400 less than you paid for the stock.

100	shares		100	shares		$600	new stock worth
x	$10	price per share	x	$6	price per share	−$1,000	original stock worth
$1,000	original stock worth		$600	new stock worth		−$400	your loss

You lost $400 of your original investment.

What fraction of the original price is $6 a share? Can you reduce this fraction to its simplest form?

24

Prices change very quickly in the stock market. The floor is often filled with people running and shouting to get information to others quickly, like these traders at the Chicago Board of Trade.

Buying and Selling Stocks

Now that you know more about how to keep track of your stock, you might want to know how to buy some shares. People investing in the stock market often make their purchases through **brokers**. Brokers communicate with the people on the floor of the stock exchange called **floor traders**. Floor traders **monitor** the changes in stock prices and finalize buying and selling. Different traders deal with different stocks.

If you ask your broker to purchase shares of a stock at a certain price per share, he takes this price to the trader of that stock. The trader monitors the stock for a few minutes on a special computer screen. If the stock's price is increasing quickly, your price may not be high enough. If the deal is completed, you owe your broker a **commission**, or percentage of that trade's value. This is how brokers earn their money.

If you pay your broker $93 for the purchase of 155 shares at $15 per share, what percent of the stock's value is his commission? To find the answer, first calculate the total cost of the stock. Next, find the decimal form by dividing the broker's fee by the total cost of the stock. Last, change the decimal into a percent.

155 shares			
x $15 price per share			
775			100
+ 155	2,325) 93.00	.04	x .04
$2,325 total value of stock	− 93 00		4.00
	00 00		
$93 broker's commission			
$2,325 total value of stock			

Your broker gets a 4% commission on this trade.

Stock Market Crashes

A stock market crash is a huge drop in the value of a large number of stocks at the same time. This happened to the American stock market twice in the twentieth century. The first great stock market crash occurred on October 29, 1929. The second occurred on October 19, 1987. Both crashes cost investors billions of dollars.

Crashes happen for many reasons, some of which are still unknown. Investing in the stock market means taking risks. When investors sense a problem or hear that stock prices are falling, they may rush to sell their stocks to try to recover the money they have invested. This causes panic and loss of faith in the market. When everyone tries to sell at the same time, there are more sellers than buyers. As mentioned earlier, if there is no public demand, prices fall because people will not pay for something that they do not want. Therefore, when the supply of stocks exceeds the demand for them, prices continue to drop and a crash occurs.

After the stock market crashed in 1929, thousands of panicked investors flocked to Wall Street in New York City, hoping to sell their shares and recover some of their money.

Investing in History

Just as people now buy and sell in the stock market, people as far back as 5,000 years ago once **bartered** and traded. The modern-day stock market is not so unlike markets of ancient civilizations. The people of Egypt and southwestern Asia discovered ways to trade with each other and other civilizations to obtain goods and resources to clothe their families, build their homes, cook their food, and improve their lives. These people often engaged in what was called "silent trade." People who did not speak the same language used different signals and sounds in order to refuse or accept the item that the other party was offering in the trade.

In the stock market today, many floor traders use hand signals to communicate with other people on the floor. The floor is generally very noisy, and signals are less likely to be misunderstood than shouted words. Also, hand signals relay information almost instantly without traders having to move long distances. It is important for traders to communicate quickly because prices change so fast. The trader who uses signals also does not announce information about an investor who may want to remain unidentified. Rather, only basic information about buying stock is conveyed. Everyone on the exchange floor must know these signals very well. A wrong signal can result in the loss of large amounts of money.

Although exchange floors certainly are not silent, traders like this man in Milan, Italy, use signals to communicate.

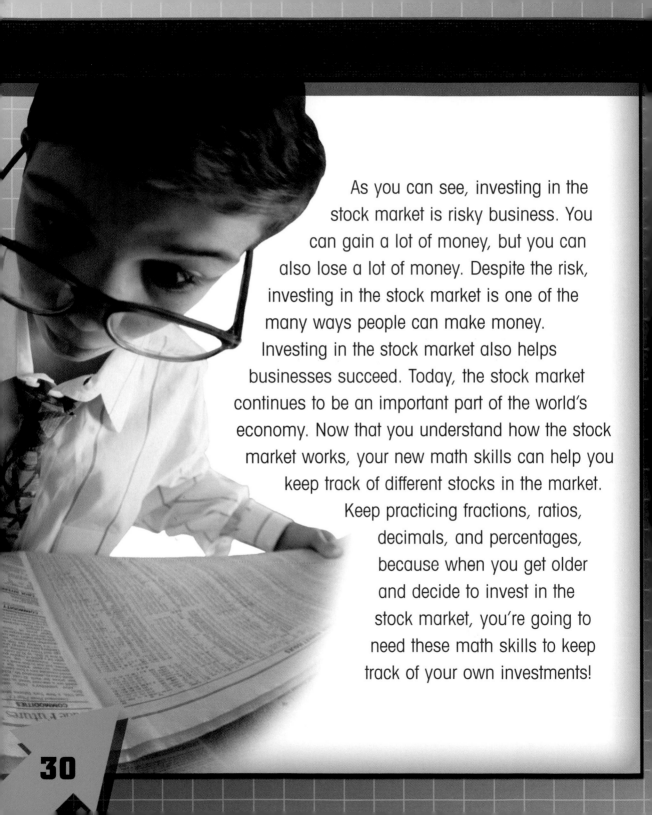

As you can see, investing in the stock market is risky business. You can gain a lot of money, but you can also lose a lot of money. Despite the risk, investing in the stock market is one of the many ways people can make money. Investing in the stock market also helps businesses succeed. Today, the stock market continues to be an important part of the world's economy. Now that you understand how the stock market works, your new math skills can help you keep track of different stocks in the market. Keep practicing fractions, ratios, decimals, and percentages, because when you get older and decide to invest in the stock market, you're going to need these math skills to keep track of your own investments!

Glossary

barter (BAHR-tuhr) To trade by exchanging one good for another.

broker (BROH-kuhr) A person who acts as an agent for others in buying and selling stock.

commission (kuh-MIH-shun) The percent of a total sale paid to a broker.

compute (kuhm-PYOOT) To figure out mathematically.

decimal (DEH-suh-muhl) A number other than a whole number that is written using a decimal point.

demand (dih-MAND) The want or need of something and the willingness to pay for it.

denominator (dih-NAH-muh-nay-tuhr) The bottom number in a fraction.

floor trader (FLOR TRAY-duhr) A person who works on the floor of a stock market and helps finalize the buying and selling of stocks.

invest (in-VEST) To put money into a business in order to make money.

monitor (MAH-nuh-tuhr) To watch closely.

network (NET-wuhrk) A large group or system of things that are connected to each other and affect each other.

numerator (NOO-muh-ray-tuhr) The top number in a fraction.

ratio (RAY-shoh) A comparison of numbers that represent different amounts.

technology (tek-NAH-luh-jee) Computers, electronics, and the Internet.

Index